Arthur Coe Spencer

The Geology of Massanutten Mountain in Virginia

Arthur Coe Spencer

The Geology of Massanutten Mountain in Virginia

ISBN/EAN: 9783337289133

Printed in Europe, USA, Canada, Australia, Japan

Cover: Foto ©berggeist007 / pixelio.de

More available books at **www.hansebooks.com**

THE GEOLOGY

of

MASSANUTTEN MOUNTAIN

in

VIRGINIA.

A thesis presented to the Board of University Studies at Johns Hopkins University, for the Degree of Doctor of Philosophy.

by

Arthur Coe Spencer.

May 1896.

PREFACE.

The study of the Massanutten Mountain was suggested by Mr. Bailey Willis, to whose personal and cordial direction both of the field work and the discussion of the data secured, the writer is greatly indebted.   Thanks are also due to Professor Van Hise, of the University of Wisconsin, for aid upon certain structural problems; and especially to Professor Clark for kindly criticism and supervision, during the preparation of the manuscript.

The recent work of Mr. Willis upon the Mechanics of Appalachian Structure, and that of Mr. Keith upon The Geology of the Catoctin Belt, have been constantly consulted, and it will be a source of gratification, if the present paper shall be judged to be a worthy sequel to the able discussion of the earlier geological history of Northern Appalachian Virginia by Mr. Keith.

Geological History

MAPS and DIAGRAMS.

INTRODUCTION.

The object of this paper is to present the lithol-
ogy, sequence and structure of the formations occurring
in Massanutten Mountain Virginia, and from their study to
indicate in so far as is possible, the processes and conditions under which the sediments were deposited, and
those by which they have reached their present attitude
and distribution.

In Appalachian Virginia there are recognized three
marked topographic belts: the Blue Ridge, the Great Valley and the Alleghany mountains.

The rocks of the Blue Ridge are largely pre-Cambrian
eruptives, but along its western slope sandstones and
shales are exposed which have been determined on fossil
evidence to belong to the lower Cambrian.   These formations are followed by a massive limestone which is characteristic of the Great Valley and this in turn by the
remainder of the Paleozoic section in the Alleghany moun-

tains to the west. In the Alleghany region folding of

the Appalachian type is made apparent by the repeated

outcrop of certain resistant strata which have given rise

to a system of parallel linear ridges with a common N. E.

and S. W. trend. The whole region has been long subject

to erosion and its present physiography serves well to

illustrate the intimate dependence of topographic form

upon stratigraphy and structure. Thus the position of

the wide valley has been determined by the presence of

the broad belt of soluble limestone, while the limiting

mountain ranges and occasional valley ridges are higher

because their component rocks have offered greater resist-

ance to the existing condition of erosion

Especially marked throughout the region is the preva-

lence of horizontal crest lines; and the fact that the

summits of the ridges are of approximately the same height

indicates an old baselevel surface — the Alleghany pene-

plain. The presence of a few low monadnocks above this

general surface, which is in all probability to be cor-

related with the Cretaceous peneplain of Professor Davis,

shows how far toward completion baseleveling had pro-

gressed before the initiation of a new cycle of erosion

by elevation of the land. The floor of the Great Valley

represents a later period of baseleveling of sufficient

duration for the complete reduction of the limestone and

shale formations without sensibly affecting the areas of

more resistant rock. Into this Shenandoah peneplain the

modern drainage has eroded its channels to a depth of 200

feet or more.

Massanutten Mountain owes its preservation to a deep

sag in the long central syncline of the Great Valley by

which a heavy sandstone stratum has been brought below the

level of the Alleghany peneplain. Subsequent erosion has

excavated the double valley of the Shenandoah on either

side and left the mountain as a residual mass forming a

narrow partition in the Great Valley.

Comprising in its component strata the greater part

of the middle Paleozoic section and isolated as it is by

broad bands of the Shenandoah or Valley limestone the area

forms a geologic unit and has been so considered in the

discussion which follows.

More sharply defined, the Massanutten area occupies the territory bounded by the two parallel branches of the Shenandoah river between the latitude of Strasburg and that of Harrisonburg Virginia.   The area is about forty-five miles long and from six to eight miles wide.

The Mountain is a system of parallel ridges which coalesce at intervals to form a chain like series of enclosed basins.   If the mountain be divided into two parts at the New Market gap South Massanutten is found to comprise three links of the basin chain while the fourth known as the Fort or Fort Valley occupies the central part of North Massanutten throughout its length of twenty-five miles.

While the mountain has, in common with the surrounding topographic features, a general northeast and southwest trend there is a slight change in direction at the New Market gap, the axis of the northern part running N. 40° E.

The general location of the area will be seen from the accompanying sketch map.

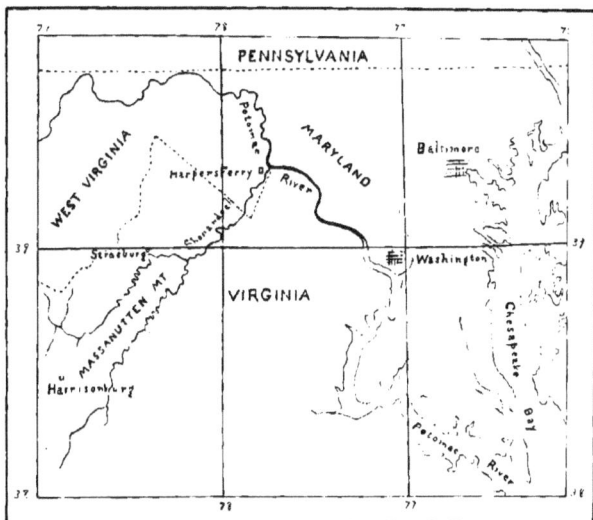

Sketch-map showing location of the Massanutten area.

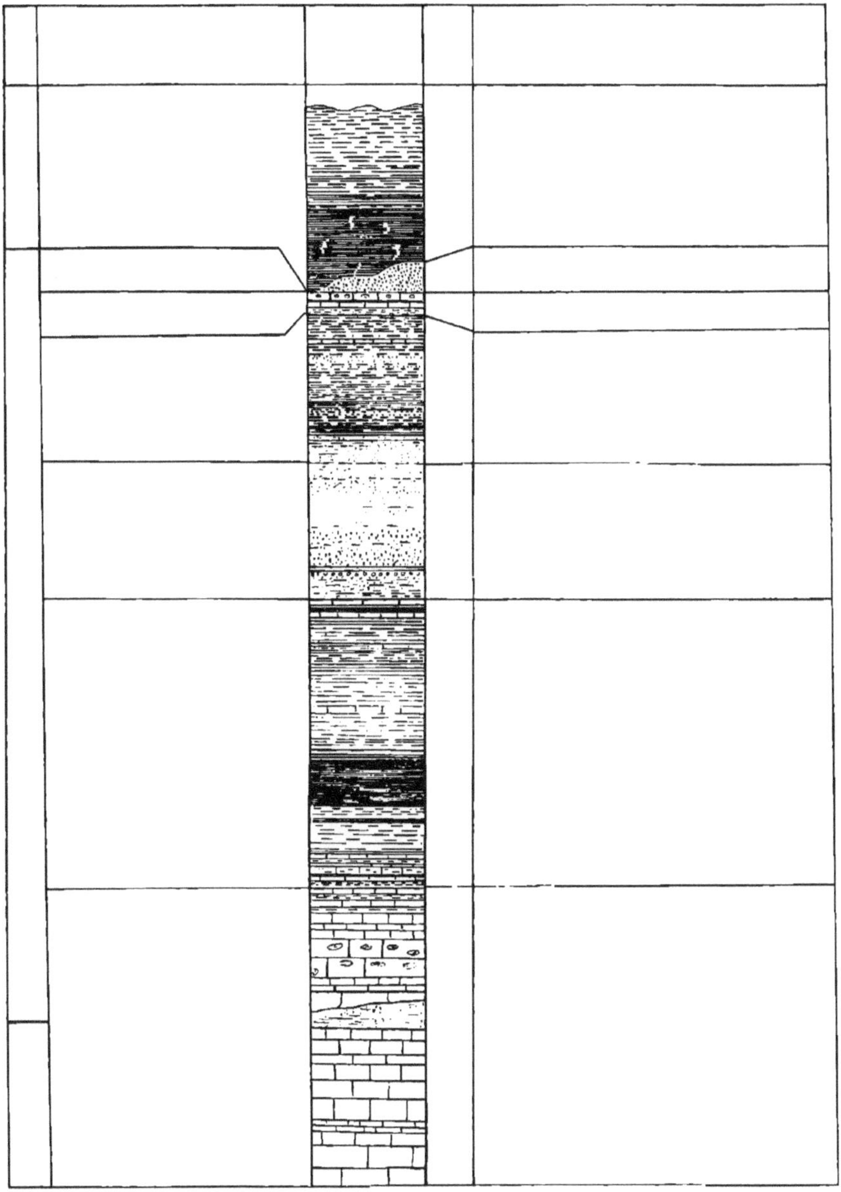

STRATIGRAPHY.

THE SHENANDOAH LIMESTONE.[1] The lowest horizon in the
Massanutten area is the Shenandoah limestone. While the
formation is typically a heavily bedded, noncrystalline,
dolomitic limestone, shaly beds are by no means rare in
its lower portion. Prominent intercalations of sandstone
are also found near the middle of the series and are known
to be particularly well developed along the eastern side of
the Shenandoah valley.[2] Similar beds of rather coarse
yellow sandstone are to be seen west of Massanutten Mountain in a railroad cutting between Edinburg and Mt. Jackson.

---

[1] The names here employed for the different formations
of the Massanutten area are those adopted by the U. S.
Geological Survey. See a paper entitled Notes on the
Stratigraphy of a portion of Central Appalachian Virginia. N. H. Darton, Am. Geol. vol. X   pp 10 - 18.

[2] A. Keith.   Harpers Ferry Folio. U. S. Geol. Survey
1894.

Mr. Keith has made the suggestion that those beds are probably the stratigraphic expression of the disturbance of environment otherwise recognized by a difference in the
faunas of the middle and upper portions of the limestone
series.[1]

Above the sandstone there are limestones containing
extensive beds of chert and above these are the uppermost
member, about 200 feet in thickness, consists of very dark,
compact limestone becoming thinner bedded and alternating
with shale to which it at last gives place.

Though fossils are by no means common in the lower
portion of the formation Mr. Walcott has succeeded in
determining their age as lower and middle Cambrian,[2] while
the upper part is Ordovician and has long been recognized
as in part equivalent to the Trenton limestone of New
York.

---

[2] C. D. Walcott. A. J. S. 1892, vol. XLIV. p. 53 and
476.

[1] A. Keith. Geology of the Catoctin Belt. 14 th.
An. Rep. U. S. Geol. Survey, 1893. p. 338.

The thickness of the Shenandoah has been variously estimated at from 2500 - 3500 feet. The sandy horizon is said to be about 1500 feet below the top of the Massanutten formation.[1]

[1] N. H. Darton. Staunton Folio, N. S. Geol. Survey 1894.

THE MARTINSBURG SHALE.   There is no sharp break between the Shenandoah limestone and the Martinsburg shale
which follows it.   Above the zone of establishment the
lowest deposits there are fine argillaceous shales which,
when fresh are of a dark drab here, but under weathering
become more or less ocherous;   they pass upward into cal-
erous shales which are frequently micaceous and often show
ripple marks and current bedding, the latter, however, on
a very small scale .   Occurring with these shales there
are bands of a somewhat coarser rock which the microscope
shows to be made up principally of angular grains of quartz
and fragments and splinters of some substance which behaves
in polarized light as a fine crystalline aggregate.   Some
calcite and a few mica flakes are always present and in one
slice a few grains of striated feldspar were observed.   An
exceptionally coarse specimen contained small lamella particles, apparently fragments from a very fine grained
shale.   The crystalline aggregate resists both hot and
cold hydrochloric acid, it is also very hard, and comparison with chert from the Shenandoah limestone leaves little
doubt that it is also chert of like derivation.   In this
rock the size of the grains is quite uniform with a max-

imum diameter of about 0.5 min.

Throughout the Massanutten area these beds are easily distinguished by their olive color, peculiar texture and somewhat prominent outcrops. Indeed they mark the only horizon within the series which can be definitely recognized.

In a railroad cutting near the bridge where the Strasburg branch of the Richmond and Danville R. R. crosses the North Fork the beds which overlie those last described are found to be very fine grained carbonaceous shale. Other good exposures of this member are to be seen at the bridge over North Fork east of Woodstock and again on the east side of the mountain on the New Market-Luray road. This shale merges into a more calcareous shale which is somewhat fossiliferous, and sometimes carries occasional thin beds of hard, earthy limestone. The increasing amount of lime in the upper part of the formation culminates in a band of very fossiliferous blue limestone which though very persistent is only a few inches in thickness. This band of limestone is made the top of the Martinsburg formation for the sediments which follow it are an expression

of the beginning of a continental movement which was to unlock the immense store of insoluable detritus which had been so long accumulating upon the neighboring land surface.

Regarding the thickness of the Martinsburg no accurate measurement was possible but the most reliable estimates place it in the neighborhood of 2800 feet.

THE MASSANUTTEN SANDSTONE. The Massanutten sandstone following the Martinsburg shale is divisible into
two members which are separated by a parting of shale.

The lower member consists at the bottom, where it
rests upon the limestone cap of the previous formation,
of very fine clay but there is a constant increase in size
of grain until shale is replaced by micaceous sandy layers
and these by coarser sandstone carrying quantities of chert
and a smaller proportion of quartz, both in rounded and
subangular pebbles.

The middle and upper parts of this series is usually
fossiliferous and except in the uppermost conglomeratic
portions is not silicified as is the case with the sandstone which lies above it.

Exposures are not frequent nor are they generally
satisfactory, for the outcrop is often covered by talus
from the cliffs of the overlying quartztile, but the chert
beds were observed at the southeast end of Short Mountain,
at the gap in North Middle Mountain and in Findley's gap

in South Massanutten. Careful search along Buzzard's Roost, at the northeastern end of the mountain, where the fossiliferous sandstone is well exposed, failed to show any fragments of chert.

The usual thickness of the lower member is about 200 feet.

The shale which separates the two members of the Massanutten sandstone is fine grained and argillaceous. Its thickness is from ten to twenty feet.

The upper Massanutten is an indurated sandstone composed entirely of well washed and sorted quartz in which cross bedding is characteristic throughout.

The greater part of the formation is a sandstone of rather coarse grain, but coarser conglomerates in which the diameter of the pebbles may be above one, and even reach two centimeters are of frequent occurrence. In these conglomerates the larger fragments are inbedded in a matrix of finer sand.

Pink and reddish layers occur locally in which the cement is in part ferruginous but almost always the bond is silicious. In some of the thin slices prepared secondary growth of well rounded quartz grains is well shown and

it seems fair to conclude that such enlargement has been the general cause of quartzitization, since the sandstone consists throughout of closely interlocking individuals of quartz. Induration has been so complete that even in the coarser grained rocks fracture surfaces pass through pebble and matrix alike without deviation, but in weathered surfaces the coarser fragments stand out in relief.

Considered as to horizontal distribution the amount of coarse material diminishes both to the northeast and southwest from the central portion of the region. The absence of coarse grains is most marked at the Peaked Mountain in South Massanutten where only a single bed was found in which the grains reached an average size of two milimeters.

In the lower part of the formation sandstone and conglomerate are interbedded but toward the top the amount of coarse material decreases and the uppermost beds are invariably fine grained. The thickness of the upper Massanutten not constant, varying from 500 feet to 800 feet.

THE ROCKWOOD FORMATION.[1] The lower beds of the Rockwood formation are sandstone but they are at once recognized as distinct from the Massanutten sandstones by their
deep red color, and where the two are seen in actual contact the line of demarkation is seen to be extremely sharp.
Closer examination shows that these red sandstones contain
a considerable amount of mica in small flakes. The red
color is due to interstitial ferruginous material as shown
by microscopic examination.

Above, the sandstone is followed by, and probably
grades into red and gray shales often rich enough in iron
to constitute a valuable ore which was mined in the region
for many years and reduced in local furnaces. No study
of these interesting ores has been attempted and indeed they
are now inaccessible except to the prospector with pick and
shovel. The so called fossil ores of New York, which are
of the same age, are thought by Foerste to have been derived
from limestone by replacement.[2]

---

[1]Named from Rockwood Tenn. See Kingston Folio U. S.
Geol. Survey, 1894.

Following the ore bearing clays and approximately 200 feet above the base of the formation ten feet of yellowish quartzite were observed at one locality.   Such an indurated sandstone but of greater thickness is characteristic of the upper Rockwood formation as it occurs west of Staunton Virginia[1] from which it may be argued that the quartzite is probably of wider distribution in the Massanutten area thann would be inferred from the single known outcrop.

Above the quartzite comes more red shale which is very fine grained and free from grit but passes upwards at least locally, into sandy shale and thin bedded sandstone.   The thickness of the Rockwood formation is from 800 - 1000 feet.

[1] N. H. Darton   Staunton Folio U. S. Geol.Survey 1894.

The following section of the upper beds of the Rockwood formation, is well exposed at Seven Fountains:

Sandstone, red, weathered------------------- 4 6

Shale, fissile, yellow,--------------------- 10

Sandstone, yellow to white, crinoid stems--

on upper surface.--------------------------- 0 6

Clay shale, fissile, yellow----------------- 6

Limestone, weathering yellow and shaly------ 1

Shale, sandy, red.-------------------------- 2

Sandstone, limy, gray----------------------- 0 6

Shale, red, sandy not fissile, about--------15 0

Sandstone hard, red------------------------- 0 9

Total exposed below Lewistown limestone-----39 3

LEWISTOWN LIMESTONE. Next in order above the Rockwood formation are the limestones and limy shales of the
Lewistown limestones. Only a single satisfactory section
is known though the distribution of the formation is to be
made out from occasional small outcrops and the more frequent occurrence of chert fragments in the soil.

In North Massanutten the best outcrops are at Seven
Fountains, on the Edinburg road and in the beds of the
streams on either side of Middle Mountain. In South Massanutten there is a small exposure near Harshbergers gap but
the limestone doubtless occurs in the Cub Run basin since
the younger shales are then present.

The limestone is full of fossil coral and brachiopods
and the microscope revealed the presence of sponge spicules
and an undoubted foraminifer of the nodosarian type in the
chert.

The following section is to be seen near Seven Fountains:

                                                        feet

Dark fissile shale (Romney shale)

Coarse conglomerate (Monterey sandstone?)-------- 1

Shale, yellow----------------------------------- 6

Limestone fossiliferous blue, beds 6 in. to 3 ft.-25

Limestone gray, earthy-------------------------- 2

Limestone, shaley much chert-------------------- 20

Limestone compact, gray------------------------- 10

Clay shale, variegate, limy--------------------- 30
                                                                     93

In the west side of the Fort a greater thickness of
limestone is to be made out and the average figure is probably not far from 150 feet.

MONTEREY SANDSTONE. The thin band of conglomerate
occurring at the top of the Lewiston limestone in the Seven
Fountains section is supposed to represent the Monterey
sandstone which is the next formation in the normal sequence.[1]
A similar, but much thicker, rock occurs along the bottom
of Little Fort valley where it contains a valuable deposit
of manganese oxide. The only other locality at present
known is on the west side of the Big Fort near the Woodstock road where a "thin bed of gray sandstone" was observed
by the late Mr. Geiger.[2]

As observed in the abandoned pits and tunnels of the Pow-
ells Fort mine, the well crystallized ore appears to occur
in a vein-like body along the bedding of the sandstone
which here dips about 45 degrees to the northwest.

---

[1] N. H. Darton Staunton Folio N. S. Geol. Survey 1894.

[2] H. R. Geiger, Manuscript notes to which I have had
access through the courtesy of the U. S. Geological
Survey.

No clay is present with the ore as is the case in the deposits along the Blue Ridge, but it is found coating the walls of the crevice in which it occurs and often as a cement binding fragments of conglomerate and penetrating the contiguous sandstone. The width of the vein varies from two feet to nine feet with occasional pinches where it is cut out entirely.

The mine shaft is said to be 110 feet in depth and from which the thickness of the sandstone must be at least 100 feet though the bottom is not reached at this place.

The great variation in thickness is to be accounted the result of erosion at a date prior to the deposition of the black shales of the next formation.

ROMNEY SHALE. The Romney shale rests discordantly upon t eroded surface of the next older formations. In its lower portions this formation consists of black argillaceous shales passing upward into calcareous shale which when exposed to the atmosphere gives rise by weathering to ellipsoidal masses of various size. These shales were not observed to be fossiliferous and no accurate measurement of their thickness has

been possible but it is certainl not over 700 feet.

JENNINGS SHALE.   The black shales of the Romney formation
are followed by dull greenish somewhat sandy shales which are
every where very fossiliferous.   There is no physical break
between the two formations and their separation is here made
on the grounds of thickness, since the normal thickness of the
basal Devonian shales in Virginia is only 600 feet.[']

A few fossils collected from this horizon were sent for
determination to Mr. Charles Schuchert of the National Museum
who writes: "The fossils you sent me are not very good for
stratigraphic determination.   However, the presence of Spir-
ifera mucronatus, Chonetes scitula and particularly Dalmanites
boothi var. galliteles, point to a Hamilton rather than to a Chemung
fauna."   In the Massanutten area evidence is not in favor of
any break between the two shales.

The shales which have been referred to the Jennings
formation have a thickness of about 600 feet.

RÉSUMÉ.   The sediments of Massanutten Mountain represent the
geological record from late Cambrian to late Devonian time.

[']N. H. Darton. Stratigraphy of Appalachian Virginia, Am.
Geologist. vol. 10. 1892 p.17.

The basal limestone, over 2500 feet in thickness, is of Cambrian and Ordovician age. It is followed by a thick calcareous shale, also Ordovician, which grades upward through an
unsorted series into a clean sandstone or conglomerate, and
above this are finer grained red deposits which give place
to a fossiliferous limestone constituting the uppermost member
of the Silurian. The basal sandstone member of the Devonian
is not always present because of erosion antedating the deposition of the black shales of Hamilton age, which rest discordantly upon the lower formations. These last, with a few hundred feet of lighter colored shales above them are the highest
beds present in the area.

STRUCTURE.

THE GREAT SYNCLINE.   Along the axis of the Shenandoah valley there is a great synclinal fold extending from
Staunton Virginia to the Potomac river and across Maryland
into Pennsylvania.   The course of this fold is marked by
a band of younger rocks replacing the characteristic limestone of The Valley.   These younger rocks, already enumer-
ated and described, attain their greatest developement in
the Massanutten Mountain and here the great flexure is
found to be a synclironum of the normal type, i. e. one in
which the axial planes of the minor folds are all inclined
toward the concavity of the master fold.   In such a fold
the axial planes converge upward.

AREAL GEOLOGY.   The double contact of the Shenandoah
and Martinsburg formations is marked in a general way by
the rivers which bound the Massanutten area, but these
streams in their meandering courses cross and recross the
line flowing alternately upon limestone and shale.

Within these limits the shale occurs in two parallel zones separated by the younger formations but coming together around both ends of the mountain and also having a narrow connection at the New Market gap. The location of several small independent areas are given on the accompanying map.

The Massanutten sandstone outcrops on all the mountain ridges forming the rims of all drainage basins and separating them one from another so that its outcrop is continuous throughout the area with the exception of that on Short Mountain and the slight break at the New Market gap.

The Rockwood formation encircles the larger valleys and forms the floor in most of the minor basins, its presence being shown in the absence of good outcrops, by fragments of red sandstone and blocks of iron ore or indurated red shale.

The known distribution of the Lewistown limestone is confined to the Fort Valley and the most southern link in the basin chain. In the former outcrops and chert fragments are frequent but in the latter only one exposure is

known. From its occurrence at either extreme of the region
it seems probable that the formation is also present in the
intermediate or Cub Run basin.

The lower of the two shale formations is found in the
principal basins, and it is possible that the upper or
Jennings shale may be present in all three but it has been
studied only in Fort Valley where it lies in two bands
separated and flanked by parallel belts of the older Romney
shale.

To recapitulate, the formations above the Massanutten
sandstone are disposed in independent areas separated by
sandstone outcrops and very noticably coincident with the
several topographic basins of the region. It is seen,
then, that these basins are primarily structural in origin.

STRIKE and DIP. In North Massanutten the strike varies only slightly from N. 40 E. except near the ends of
structural basins when, parallel outcrops coming together
the strike sweeps through 180 . In South Massanutten corres-
ponding with its more northerly trend the average strike is
N. 30 E. Throughout the whole region the dips are as a

rule steep and frequently reversed, but they are not pre-
vailingly away from the central axis except for limited
distances in the Martinsburg outcrops and along the bound-
ing ridges.

PITCH.   The origin of the several structural basins
is due to undulation along the axis of the great syncline.
Each basin is separated from its neighbor by a node or
point of inflexion where the mountain has a diminished
width.   Approaching these nodes the hitherto parallel outcrops converge and the massive sandstone stratum, rising
from one depression changes its pitch and passes down into
another.   The most marked node is at the New Market gap
where the mountain has been pinched in two by the failure
of the Massanutten sandstone. (See the longitudinal section
on plate 2.)

Conspicuous variations of pitch in the secondary folds
of the syncline will be considered under the head of minor
folds.

MINOR FOLDS.   The amount of folding which the different rocks have undergone is by no means constant: the
massive Massanutten sandstone has been thrown into rather

wide folds which are never closely appressed as is the case
in the thick shales of the Martinsburg formation. For
this reason and because its outcrops are always prominent
and the attitude of its beds unmistakable, the Massanutten
sandstone is better suited than any other formation for a
study of second order folds and their relations to the
great syncline.

The folds which have been determined by the massive
sandstone are from three-fourths to two miles in width
and of variable length. The longest which can be definitely traced is the Little Fort syncline, thirteen miles
from end to end.

The individual folds are of uniform width and their
limited length, as in the larger flexures, is the result
of pitching axes.

Where exposed in cross-section, as in the case of
anticlines cut through by streams, they are often seen to
bear still smaller flutings, constituting folds of a third
order, with their radii shorter than the thickness of the
formation. It seems possible that these corrugations are

related to the second order folds as they are in turn to
the great syncline though it has not been possible to make
sufficient observations to show that this is the case.

In horizontal plan the axes are parallel in any given
latitude, but overlap en echelon bringing the nodal point
of one axis opposite an inter node of the adjacent fold, so
there seems to be a tendency towards alternation of
synclines and anticlines along the axes of folding which,
of course, have the same general trend as the rock strike.

Considering now the general relations of the folds, it
is found that they are symmetrically arranged with respect
to the master axis.   In general any section across the
mountain will show the same number of ridges on either side
of the interior valley, and in those equally placed as to
number and distance from the centre there is correspondence
of phase: for instance in the central part of North Massanutten there are two ridges on either side, the inner are
both anticlinal and the outer are both on the rising limbs of
the succeeding synclines.

Near nodal points the attitude of the folds is usually upright symmetrical but the longer corrugations are, as a rule, overturned in their central portions. The series of sections given in the accompanying figure (plate 2.) represents the structure of the Massanutten sandstone alone and being drawn to scale gives an idea of the proportion between the thickness of the stratum and the dimensions of the folds.

Lack of definite horizons has made it impossible to determine the general laws of structure in the Martinsburg formation. In no instance has it been possible, from the most careful field notes, to construct a section which could be considered reliable. Frequent changes in the direction of dip with much rarer differences in strike indicate flexures of a high order and in the absence of beds easily recognizable together with more than occasional breaks in continuity of outcrop, preclude the possibility of determining the position or presence of broader features comparable to the second order folds of the overlying sandstone.

It is in accord with the prevailing ideas of the be-
havior under deformation, of rocks differing in composition
and texture, that there should be folding in the thick,
homogeneous, and by no means strong shale which is not
shared by the more resistant rocks above and below it.
Evidence is not wanting that independent crumpling has
taken place, but it is worthy of remark that there is always close agreement in structure between the upper beds of
the shale and the sandstone above it, and between the lower
beds and the underlying limestone.   Whereas in the section
from Maurertown to Seven Fountains there are only six folds
of the Massanutten sandstone in the four and a half miles
between its limits, no less than seven anticlinal axes are
apparent in the band of underlying shale about two miles in
width from its contact with the limestone east of Maurertown to the bottom of the sandstone in the mountain ridge.
On the east side of the area there is an equal width of
shale exposed and in a section along the river west of
Rileyville eight anticlines are to be made out.   In this
section the limestone at the base is overturned and though
the lowest beds of the shale are not exposed the sandy member is soon seen dipping 70 to the southeast and continu-

ing to reappear in isoclinal folds nearly to the bend in the river where the folds become more open and the sandy beds pass beneath the more calcareous beds above. This close folding is undoubtedly peculiar to the shale and not shared by the Shenandoah limestone. Another section illustrating the greater amount of folding in the Martinsburg is the one between Strasburg and Riverton, where the positions of more than twenty folds have been determined, though this figure does not represent their total number since observations are wanting over parts of the section.

CLEAVAGE and JOINTING. In the homogeneous shales throu out the entire region an imperfect cleavage h s been developed and frequently is a source of difficulty in ascertaining the true strike and dip of stratification planes for which it may, at times, be mistaken. Wherever observed the cleavage strike was found to cross that of the rocks at an angle of about 20 , its usual course being N. 20 E., but locally it was noted as due 1 north. Similar discrepance was observed by Rogers many years ago.

34

He says: "Cleavage in the southeastern Appalachian region exhibits a steep Southeasternly dip; but it is interesting to note that the strike of the cleavage seems not to coincide with either the strike of the Paleozoic strata to which it belongs, nor yet with the strike of the previously disturbed and somewhat syenitic gneiss." The prevailing dip of the cleavage on both sides of the mountain is to the east and it is usually steep, ranging from 60 to 80 with the horizontal.

When jointing is to be made out there are always two systems nearly at right angles and the joint planes are usually nearly vertical. At Strasburg the principal jointing has a direction N. 10 E. dipping to the northeast 85, while the dip in the second system normal to it is 55 to the southeast. At Edinburg the first jointing is N. 75 E. with the dip southeast 85.

35

FAULTING. Below the eastern crest of North Massanutten extending from Kennedy's Peak for about ten miles to the northeast, there is a second ledge of sandstone which has been cut into a series of knobs by the waters of frequent springs. The sandstone rises in a sheet, apparently out of the underlying shale, to a height of over 100 feet, then continues more gradually with the exposure of the ragged edges of the strata, to a somewhat greater elevation, here there is a bench free from sandstone blocks and probably underlaid by a shaly rock though no outcrops are known. Beyond this terrace the sandstone again rises, this time from 200 - 400 feet, often like the lower ledge slightly overturned to the west, but with essentially the same dip, except at Kennedy's Peak where the inclination in the higher ledge is very much less than in the lower.   In the lower outcrop there is much evidence of dynamic action in very perfect slickensides.

To consider these two sandstones conformable would require a total thickness of over 1200 feet, about twice the usual measure of the formation.   Also the open syncline given by Rogers in his section of Massanutten is not in

accord with the facts, since the first depression west of
the high ridge is not anticlinal as represented, but an
overturned syncline.   There remain two alternatives: firs
the two sandstones are the two limbs of a very close syncline; second, there has been dislocation.   If there is
a close synclinal fold the summit of the ridge on the oute
limb should not, in the light of later physiographic development, be lower than that on the inner limb.   Any ordinary fault would be open to the same objection but the
feature has not been studied with the minuteness that will
warrant the assumption of complicated displacement of the
strata, so that its structure is still an open question.

A somewhat similar problem is presented by the break
in continuity of the sandstone in Middle Mountain (North
Massanutten) where the Martinsburg shale is brought to the
surface at an elevation of 1900 feet by a pitching anticline.   The rising limb of sandstone forms a ridge above
the 2000 foot contour on the east, but on the west the
first sandstone is exposed in an inconspicuous ridge some
200 feet lower.   The only explanation which has suggested

itself is that of a stretch fault, by which the sandstone
was pulled away to allow the rise of the Middle Mountain
arch.

RÉSUMÉ. The Massanutten syncline is complex flexure in-
volving a thick body of Paleozoic sediments consisting of
a massive limestone about 3000 feet in thickness, and this
by a massive though thinner quartzite sandstone above which
there is present a considerable thickness of fine grained
deposits consisting of limestones and limy shales.

The trend of the great fold is NE. and SW. and it is
separated into several structural basins by reason of variation in axial pitch which also determines the length of
the subordinate folds.   These folds are most evident in
the structure of the sandstone stratum where they are easily followed as rather open corrugations often overturned
toward the center of the system and constituting the great
fold a normal synclinorium.   Even the basal limestone is
sometimes overturned toward the mountain and in the shale
upon the east side of the area isoclinal folding occurs
with its dip to the southeast.   It cannot be said, however,
that the prevailing dip is away from the central axis of the

area though this is the case in the sandstone of the bound-
ing ridges.

Imperfect cleavage is present in all the shale formations with its steep dip usually to the east and its average strike N. 20 E.

Faulting is appealed to as a provisional explanation
of the flanking knobs of sandstone along a portion of the
eastern ridge and again in the structure at the Middle Mountain gap, but in both cases the displacement is small.

39

TOPOGRAPHY.

Having already considered the general relations of the Massanutten region and pointed out the principal topographic elements thereof, it now remains to give a more detailed discription of these physical features.

THE MOUNTAIN RIDGES. The ridges of Massanutten Mountain, capped in every case by the Massanutten sandstone, are of three types: monoclinal, synclinal and anticlinal. The general synclinal structure of the region being known it is apparent that the outer rim-forming ridges are monoclinal, while the outlying members of the system (Short Mountain, Lairds Knob spur, and its companions immediately to the east) are the keels of long synclines, and the interior ridges are anticlinal. These last comprise the ridges which separate the synclinal side valleys from the deeper Fort valley, North and South Middle Mountains, and two or three others of less importance.

Wherever two ridges unite or a pitching anticline rises to a sufficient height, knobs are formed which rise from 100

to 700 feet above the adjacent ridges.   The following tabula-
tions of elevations are taken from the topographic sheets of
the U. S. Geological Survey.   Table I. gives all the principal
peaks with the altitude of intervening ridges;   read from left
to right it shows approximately, the elevation of the parallel
ridges at the same latitude;   from top to bottom it gives the
figures along the axis from NE. to SW.   Tables II and III. give
the elevations of peaks and ridges separately.

I.

˅2000   ᶠ2200

                                                         2000

1900   1500   ×2200

                                                         1800

1800   1400   2000

1700   2000   1800

2700   †2700

2600   2300   ×2500

                            †2700

2700   2500   2100   2000

                ˅2600   ^2600

                        1900 (gpt)

                        2500   2500

                2600   2000

                2800   †3000   2500

             ×3300   2800   2600

                        2800   2600

                            ×2900

^ Synclinal peak

† Anticlinal peak

II.

          2000   2200

                          2200

          2700   2500

2700   2700

          2800   2600

                2500

                          2500

          3300   3000

                2900

III.

1900		
1800		2000
1700		1800
2300		
2500	2100	2000
	1900	
2600		2000
2800		2500
	2800	2600
	2600	2800

From these tables it is seen t.a the difference in the
elevations reached by the most resistant stratum (the Massanutten sandstone) has a maximum of fully 1600 feet, but betwee:
t'e two extremities of the mountain it is only half this figure.   The grouping into two levels is well shown in tables II.
and III. as is also the intimation that there is a rather definite distance-relation between them.   The knob at the southern
end of Little Fort is 400, 700 and 900 feet higher than the
respective ridges south, northeast and north of it, and Short
Mountain reaches a like elevation (2700 feet);   Kennedy's Peak
is 600 feet higher than the gap just below it and 500 feet a-
bove the monoclinal ridge near by;   Middle Mountain peak rises
600 feet above a similar gap and 500 feet above the succeeding
ridge.   The knobs which terminate North Massanutten are 2800

feet but across the New Market gap (elevation 1900 feet) the

peaks are only 2500 feet.   In South Massanutten the only

marked example is the long spur which connects Lairds Knob

with the main mountain; like Short Mountain, it is synclinal

and its knob is the highest point in the Massanutten system,

(3300 feet) and is 500 feet above the very uniform ridge just

east of it.   It seems remarkable that there is no marked rise

at either extremity of the mountain for though in both cases

there is a pitching syncline but the increase in elevation is

only a matter of 100 feet.   That the lower or ridge level is

a base-level surface does not admit of doubt, but there is

less certainty about the upper or knob level.   It seems very

possible that we may have here in the Massanutten Mountain the

clue for correlating, more decisively than has yet been possi-

ble, the Catoctin and Alleghany base-levels already studied in

some detail by Mr. Arthur Keith.

THE SHENANDOAH PENEPLAIN.   Viewed from a slight elevation

the floor of the Shenandoah valley appears as a broad undulating

'A. Keith   Geology of the Catoctin Belt.   14th Ann.

Rept. U. S. Geol. Survey 1895.

plain with occasional hills rising above its general level.
This surface, which may be called the Shenandoah peneplain,
has an elevation, near Strasburg and Front Royal of about 900
feet but it rises gradually toward the SW. attaining a height
of 1500 feet in the vicinity of Harrisonburg. The rocks
which have been reduced to this level are principally the limestone and shale of the Great Valley but within the Massanutten
basins it is usually also well developed in the uniform rocks
younger than the Massanutten sandstone. The latter seems to
have the only stratum capable of withstanding the eroding in-
fluences. Within the interior valleys this old erosion plain
has a higher elevation than outside, though its slope is about
the same, for in The Fort, where axial drainage makes direct
comparison possible, it rises from 1000 feet near the northeastern end, to 1300 feet just below the commencement of the
Middle Mountain partition, the distance being only twenty miles
In the Little Fort valley the divides between the small streams
which drain its central part, have an elevation of about 1700
feet, and in the narrow drainage basin of South Massanutten
there are sloping terraces above 1800 feet. These levels and
the somewhat greater elevation of the ancient floor of The

45

Fort valley, are the result of local base levels due to the resistance which the Massanutten sandstone must always have offered to stream corrasion, especially where the streams have been of small volume.

In the valleys of both branches of the Shenandoah river there are some very noticable deposits of coarse gravel and cobble stones which, though their distribution has not been studied, seem to be confined to the vicinity of the rivers. Along North Fork, the deposit is mostly made up of rounded sandstone fragments but in the other valley they contain also a large proportion of schist porphyry and amygdaloid. Here the material is also well rounded and often of large size, above one foot in diameter in the case of certain boulders of a dark blue quartzite. Near Rileyville the gravels are very well developed on both sides of the river and in the railroad cuttings south of the station are exposed to a depth of twelve or fifteen feet. At this place the deposit is over 100 feet above the bed of the river.

THE DRAINAGE. The present drainage of the Massanutten area has cut the Shenandoah base level-surface into deep intaglio and the modern streams are flowing from 200 to 300 feet

below the older level.   The courses of the principal streams
are parallel to the general trend of the S enandoah valley and
the rock-strike; thus both branches of the Shenandoah river
follow the contact of the limestone and shale and flow from
one to the other with apparent indifference.   This is true
also of Smith's creek which flows along the southern half of
the western boundary of the area under consideration.   The
course of Passage creek is also axial, as is that of the other
mountain streams until they leave the interior basins, when
they turn at right angles to join the master stream.

The two rivers are extremely crooked, which is the more no
ticeable in that their grades are corrasion rather than solution slopes.   Imperfection of stream adjustment is, moreover,
seen in the irregular distribution of grades, for the fall per
mile (reckoned between the hundred foot contours) varies from
4.5 feet to 11.7 feet.   The average fall per mile is 7.69
feet for South Fork and 8.20 feet for North Fork.   The variations in declivity are well shown in the accompanying diagram where the vertical represents fall in feet, and the hor-

izontal river-length in miles.   The slope is represented as
uniform between the hundred-foot contours.   The distance between the 500 foot and the 1000 foot contours on South Fork
is four miles greater than on North Fork.   Both rivers have
much greater slope above and below the 700 and 900 foot contours than in the interval between them, which is very nearly
equal in either case.

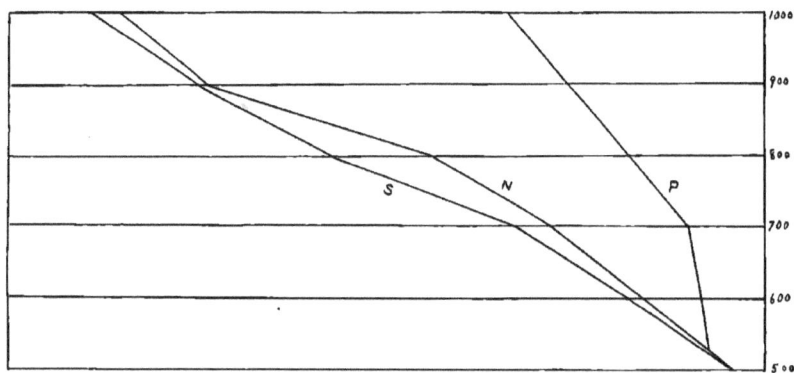

S - South Fork, N - North Fork, P. Passage Creek.

Passage creek is in one respect a remarkable stream for
unlike the ordinary stream flowing in a canoe-valley it finds
its exit through a gap of its own excavation in the very prow

of its long basin, and this in spite of the fact that at least
two wind gaps give evidence of former possible outlets through
the lateral ridges.

Besides draining the great synclinal basin of North Massanutten Passage creek also drains the larger part of the area
between the main ridges and Short Mountain. These waters it
receives through a gap in the sandstone rim which has an elevation of less than 1300 feet. Between this gap and the exit
of Passage creek there is a fall of 600 feet. This makes the
grade within The Fort about 30 feet per mile which is noticably
less than in the portion of the stream outside of the mountain,
where it is approximately 50 feet per mile. This is a good
illustration of a secondary base level determined by a resist-
and stratum.

RÉSUMÉ. In the Massanutten area topography is closely
related to stratigraphy and structure. The most resistant
rocks are found to outcrop at the greatest altitudes and the
drainage basins between the ridges to coincide with structural
basins from which the more easily eroded rocks have been removed.

The ridges and peaks, which are capped by the Massnutton
sandstone, seem to belong to separate periods of base leveling
since the former are, in general, about 500 feet higher than
the latter.   The effect of later erosion is seen in the reduction of the less durable limestones and shales to a base
level surface about 1200 feet below that of the mountain
ridges, and also by the dissection of this surface by the pres
ent streams.   Bordering both branches of the Shenandoah river
there are extensive deposits of coarse gravel which have not,
however, been investigated.   The rivers themselves are of
peculiar interest because of their meandering courses, steep
declivity and the irregular distribution of their grades.

GEOLOGICAL HISTORY.

1. DEPOSITION.

EARLY PALEOZOIC PERIOD. The interpretation of the earliest
sediments occurring in the Appalachian region of Northern Virginia and Maryland has been made the subject of careful study
by Mr. Arthur Keith.[7] In an area adjoining the Great Valley
he has shown that the sea advanced eastward, in lower Cambrian
time over a land-surface which was not, in diversity, unlike
that of the present day. This transgression resulted in the
deposition of basal deposits of variable and composite character and, conquest of the land being continued, the main current
became so regulated that complete sorting of materials was
made possible, and shales and sandstones were deposited. Fi-
nally, with continued depression, the amount of clastic sediments was reduced and the deposition of the Shenandoah limestone accomplished. This formation Mr. Keith believes to
have been spread in a continuous sheet over the present site
of the Blue Ridge.

SHENANDOAH PERIOD. The conditions which prevailed during
the formation of the Shenandoah limestones are outlined by
Mr. Keith in the following words:[1] "The cause of the change
from silicious to calcareous sediment at the beginning of
Shenandoah time receives but little elucidation from the facts
of this region (The Catoctin belt). It is obvious, however,
that the change was continental, so that the solution which
fits other areas will also be suitable here.

At first thought subsidence will be appealed to as the
cause, since limestones are more in the nature of sea than
shore deposits. In the region of the Catoctin belt subsidence was but another step in the direction already pursued....
The unavoidable inference from these facts is that with the
beginning of Shenandoah deposition came depression and submergence of considerable areas of land....Within the limestone body at some horizon came the change from Cambrian to
Silurian."

After discussing the sequence and distribution of shales
in the lower part of the Shenandoah and of the sandy and shaly
beds in the middle part, he continues: "In these series of

---

[1]Op. Cit. p. 337, et seq.

facts several general relations stand forth between organic
and inorganic systems. The lower limestones contain lower
Cambrian fossils. The medial slate and sandstone group
gives evidence of instability of environment, and a slight
reversal of the process of submergence. These rocks contain
a fauna of lower Cambrian age with other forms suggestive of
middle Cambrian. The upper limestones are defined by lower
Silurian fossils. Breaks of the organic series occur in the
slate-sandstone horizon, therefore, from lower to middle Cambrian, and from middle through upper Cambrian to Silurian.
It is natural to infer a causative connection between these
breaks and the oragraphic movement occurring at the same
time....

To sum up, the Shenandoah limestone resulted from the
perfect adjustment of erosion and sedimentation, giving a uniform product over very wide areas and lasting for a very long
time. It was initiated by depression which must have been
great to allow for its considerable thickness, 3000 feet, and
which was not simultaneous, since it was interrupted by erosion
along its eastern margin. Its land masses must have been of

considerable height to escape submergence, yet the fine sediment that they furnished shows a perfect stream adjustment
and good progress in degradation to baselevel. It marks, in
short, the culumniation of the process of depression and degradation which was initiated in Loudon (earliest Cambrian)
time."

MARTINSBURG PERIOD. The marked difference in the character
of the Shenandoah and Martinsburg formations, when account is
taken of the great mass of each, leaves no doubt that the displacement of the limestone by shales must be significant of
important changes in the pre-existing relations of sea and
land. The perfection of the passage between the two formations with continuous deposition is, however, sufficient to
show that whatever revolutions may have brought about these
changes the Massanutten area was remote from the position of
their maximum manifestation. In other words, the open sea
conditions which prevailed during Shenandoah time were not
changed to any great degree. The Massanutten phase of the
Martinsburg was then, an off-shore deposit, and as such offers
a generalized record of the events transpiring in the course

of its formation which would have been more or less hidden
by the detail of deposits formed in the littoral zone, where
every oscillation large or small, would have left its mark.
¶The features of the Martinsburg formation have been fully described in another place, but the essential points for the
interpretation of events are: at the bottom, fine clay shales
alternating with thin bands of limestone, above this, calcareous shale carrying mica often ripple marked, and intercalated
with it sandy shale containing fragments of chert derived
from the Shenandoah limestone; and lastly the greater part
of the formation which is increasingly calcareous toward the
top.

    While there are, in general, two factors which may be
instrumental in bringing the loose materials of a previously
base leveled continent under the action of the sea, i.e. cli-
matic change and continental deformation, the first, if it
was an element in the present case at all, is indeterminate,
and was at best subordinate to the second.

    Appealing then to deformation, it will be seen that an
advancing sea would, beside bringing the old land surface

under direct wave action, render those parts of the continent
in which there may have been greater relief, more available as
a source of sediment by decreasing their distance from shore.
Taken with the shallowing of the water, which is necessary
for increased transportation by marine currents, this lowering
of the land constitutes a landward tilting about an axis west
of the shore line.   The strongest evidence in favor of the
view is the occurrence of slates of probable Hudson age, far
east of the Blue Ridge in Buckingham county Virginia.[1] Mr.
Keith argues from this occurrence that the Martinsburg formation (equivalent in par with the Hudson shales) once continued over the Blue Ridge and overlapped the Shenandoah limestone.   Mr. Keith says.[2] "From the above facts it is clear
that the close of the Shenandoah limestone was caused by wide-
spread dynamic action and continental movements.   In the

---

[1] N. H. Darton. Fossils in the "Archaean" rocks of central
Piedmont Virginia.   A. J. S. 63) vol. 44. p. 50.

[2] Op. cit. p. 344.

.

region under discussion the movement was plainly tilting, elevation in Ohio and depression in eastern Virginia.   In other cases it was mainly elevation of smaller tracts along the shore."

The other cases referred to by Mr. Keith in which elevation seems to have been more local, are founded on an erosion interval between the great limestone and the shales which follow it, so that the explanation given by him for the origin of Martinsburg conditions is not in accord with the order of events clearly recognized in the areas cited.[7]   Also in the light of the Massanutten sediments it seems to be certain that there was no marked advance of the sea during Martinsburg time.

On the other hand, a retreating sea, laying bare a portion of the sea floor, would be the result of continental elevation decreasing the depth of the sea and increasing the height of the land thus bringing about, at once, more competent drainage and the possibility of stronger marine currents.

---

[7]For these references see Mr. Keith's paper p. 344.

This suggestion as to the nature of the deformation which closed Shenandoah time is supported by a study of the Martinsburg sequence, already presented.

Wherever the Shenandoah shore may have been, it is almost certain that between it and the first outcrops of the limestone now exposed there were extensive deposits of shale constituting the littoral facies of sedimentation during a long period of low relief in the adjacent lands. These shales could hardly have contained much coarse material because of the adolescent character of the existing drainage, though the possibility of some such detritus is to be recognized, since the sea may have been at times transgressing the old land surface. Also somewhere near the top of the Shenandoah series, but of unknown extent landward, there were chert beds probably not included however in the shore deposits, nor yet overlaid by them, but passing into them horizontally.

With these conditions in mind, and with the further assumption that there remained an eastward-lying land-mass covered with a deep residual mantle, the general applicability of the elevation hypothesis to the origin of the Martinsburg sediments is at once apparent.

The accession of the waves to the Shenandoah shales,
with conditions of carriage variable though slightly increased,
accounts very well for the fine clays alternating with limestone bands at the base of the formation, while the complete
establishment of shale over limestone may indicate a further
slight uplift with shallowing of the sea. Up to this time
the comminution of wave derived material was very complete and
only the very finest particles reached the Massanutten area,
but during the culmination of uplift, which is recorded in the
sandy layers, though carriage was effective, grinding was not
complete, for the very refractory chert and quartz fragments
are accompanied by flakes of dark shale and sometimes even by
feldspar.   This would doubtless indicate that the waves
were handling a large amount of debris much of which must have
been wave-derived, since the chert could have come only from
the underlying limestone, and the shale fragments from older
shales near by. The quartz and infrequent feldspar alone
seem to have been brought in by streams, though only the latter
necessarily requires this explanation, since the quartz could
easily have had its immediate origin in the rocks upon which
the waves were at work. Following out this hypothesis it is

necessary to suppose that the recession of the sea had laid
bare a broad zone of the sea floor, from which the waves had
removed probably the greater part of the argillaceous rocks
if these ever covered the chert beds — or if there was horizontal passage between the two, the amount of emergence must
have been even greater.

During the deposition of the sandy beds the sea had that
indefinitely moderate depth at which ripple works are possible,
though conditions of either supply or transportation, or both,
were not constant, for very calcareous shales are intercalated
with the sandy layers.   In the upper member of the Martinsburg, the increasing proportion of lime can only be explained
by the diminished competency of marine currents, at least in
their relation to the region under discussion, though they were
in all probability still effective farther east toward the
shore, in which case the zone of maximum deposition must have
been likewise shifted in the same direction.   A thickness of
2000 feet for these calcareous shales is evidence of the con-
tinuance of the same conditions during their deposition and it
is probable that the position of the shore was practically un-
changed throughout upper Martinsburg time.

60

What part the rivers may have taken in furnishing sediments
is not to be made out from any rocks now available; certainly
the Massanutten phase was too far from shore to give the data
required, so that, while it may be taken as proved that there
was uplift to allow of recession of the sea, it is by no
means certain that the entire continent was raised to a com-
mensurate amount. Indeed, the disturbance may have been
localized near the Shenandoah shore and the general uplift
have been moderate, though sea-ward it was certainly enough
to cause noticable decrease in depth of water.

MASSANUTTEN PERIOD. The second period of great change is
expressed in the sediments which follow the fossiliferous
capstone of the Martinsburg shale. The disturbances at this
time initiated, were to unlock the store of coarse material
which had been collected and prepared during the deposition
of the preceeding limestones and shales.

The concentration of the more resistant products of rock-
decay for the formation of extensive sandstones is a process
of great complexity. During the early part of a topographic
cycle, when denudation exceeds decomposition, the numerous

rapid streams deliver to the sea large quantities of a coarse
materials of a mixed composition which may be partially sorted
by the waves and somewhat widely distributed by marine currents; but as the drainage becomes more mature there will be
a tendency for stream derived detritus to accumulate at the
mouths of large rivers, often without much wave washing. The
deposits in either case will be more or less heterogeneous,
and the same will be true of the deposits derived by a transgressing sea from a land-surface covered with a deep residual
mantle. Such coarse grained, heterogeneous deposits would
be confined to a rather narrow littoral zone, but if a slight
elevation of the land should bring them a second time under
the action of the waves, they would be resorted, the less resistant components would be largly ground up and removed as
fine silt, and the zone of coarse deposition would be moved
seaward. The rivers would be bringing fresh material, but
with continued slow, or intermittent rise of the land accompanied by marine planation, these and the older surviving materials becoming mingled, the final result would be a mass of
sand and pebbles composed almost entirely of quartz.

In the Massanutten formation the points in evidence are the micaceous shale passing through sandy shale and sand, into the very coarse conglomerate made up of quartz and chert pebbles in a finer matrix, the sharp replacement of the conglomerate by a thin somewhat sandy shale, and the equally abrupt appearance of a sandstone, made up entirely of quartz, coarse in its lower beds but becoming finer above.

Considering first the chert pebbles, they were derived either from the zone of limestone uncovered at the beginning of Martinsburg time, or from the shore deposits of that formation. In the first case the sea must have advanced, since so large a quantity of residual chert could not have been brought together from an area so limited, by the activity of running water. In favor of this explanation is the preponderance of chert over quartz in the conglomerate below the shale parting, and its almost complete absence in the upper sandstone member; against it is the large size of the pebbles, requiring offshore currents of extraordinary power.

The second hypothesis is the more attractive, and, accepting it, the events may have been somewhat as follows:

during Martinsburg time coarse deposits were forme between

Massanutten and the shore and it seems as though there may

have been a coastal plain, not unlike that bordering our present

Atlantic, over which the sea swept from time to time covering,

in all probability, the previously exposed chert beds, and that

the earth movement which closed the shale period caused a com-

paratively sudden retreat of the sea,which brought the shore

line near to the position held at the eginning of the Mar-

tinsburg period.   At this time the bottom shales of the Mas-

sanutten formation were deposited and with continued elevation

of the land, the shore deposits formed during the erosion of

the chert were reached by the waves, and the shore was then

farther west than it had been at any time previous.   It must

have been very near the Massanutten region, for pebbles an

inch and a half in diameter could not have been brought from

any great distance.

Whatever the intermediate oscillations may have been, resumption of silicious deposition found the shore, once more not far from the area under consideration, if the conglomeratic character of the first deposits may be considered a criterion. Powerful currents must have swept the shale clean, and carried over it very much arenaceous material before deposition began. In the thick sandstone which follows, the ubiquitous cross-bedding is indicative of conflicting currents such as might result from strong tidal currents meeting a powerful under tow, but there is no evidence of varying conditions of environment. The unity of this member of the formation is in favor of the assumption that the period which it represents was one of quiescence, in which the only changes in the relation of land and sea were those brought about by marine or subaerial denudation. Under such conditions the shore facies of the preceeding formations, having been well washed in the manner already outlined, would be effectively re-worked and widely distributed.

From the greater quantity of coarse gravel in the central part of the area it seems possible that opposite the Massanutten area, during both Shenandoah and Martinsburg time, there

may have been the mout' and delta of a great river.

The nature of the movement with which the Massanutten period began is most simply and satisfactorily explained as a direct uplift without tilting, but evidence is entirely wanting to show whether it may not have been diferential, with local folding along a coastal zone; but whatever the movement may have been it produced shallowing of the sea, and these conditions were not seriously interrupted until over 800 feet of sand had been spread over a wide area.

ROCKWOOL-LEWISTOWN PERIOD. The change in color from white t red and the appearance of considerable quantities of mica has been made the basis of separating the Rockwood from the Massanutten formations, since the physical conditions under which the two were deposited must have been very different. The Rockwood formation is extremely variable, comprising sandstones, shales and limestones, but characterized by the large amount of iron which it contains. Also in the Lewistown formation, though limestone makes up the greater part, it is accompanied by sands and clays. The only differences seem to

be the greater proportion of calcareous matter and the absence
of red color, and it is impossible, at least upon physical
grounds, to separate these formations at any definite horizon.

It is very difficult to reach any conclusion as to the
conditions under which the Rockwood formation was deposited.
Any time theory must be applicable to the wide-spread formation
which carries the "fossil ore", as a whole, and for its formu-
lation would require an extended knowledge of the different
occurrences.   It may not be out of place, however, to suggest
some possible changes which may have transpired to bring about
the difference between the Rockwood and the Massanutten formations at their contact, and to attempt to show, in a general
way, the sequence of events connecting it with the next terrane
above.

If the inferences already set forth regarding the origin
of the Massanutten sandstone be accepted, the question becomes
pertinent whether the shales and limestones which follow it
may not represent the latest depositional phases of a topographic cycle initiated by elevation in early Massanutten
time, and in general, this explanation seems to be applicable.

Marine planation having continued throughout the deposi-

tion of the white sandstone it is supposable that the sea had cut its way across the old littoral deposits, possibly with the aid of a slight sinking of the land, and reached a region of deeply decomposed crystalline rocks. With the large amount of heterogeneous materials which the waves might then receive, they would not be able, with a single treatment, to efficiently sort or disintegrate. The result would be, with the strong currents of the Massanutten period still in force, the transportation and deposition of materials of mixed composition, but graded from coarse to fine with increased distance from shore. The gradual passage from sandstone to shale may represent either a gradual decrease in the material received by the waves or what is more probable a partial checking of current by slightly deepening water, perhaps accompanied by farther advance of the sea. With the cessation of clastic materials the beds which now bear the limonite ores were formed, after which land-derived detritus was again brought in, and a period of oscillation recorded in the alternation of rocks of different character. Such oscillations would give rise to a coastal plain upon which the materials for a subsequent coarse formation would be scoured and sorted. Indeed

the process had gone far toward completion for in the upper
part of the Rockwood there is a persistent though thin band
of quartzite, made up of perfectly clean quartz grains of unifor
size. This stratum must mark the greatest of the many recess-
ions of the sea, and in its constant character over the region
west of the Great Valley, denotes that the sea had a uniform
depth, both before and during its deposition. Above the white
sandstone, shales grading upward into red sandstones followed
again by calcareous shales, bear witness to continued variation
of environment, but the records are those of a rather shallow,
though deepening sea at a distance from shore.

During all this time the land has been undergoing erosion
and a slight depression of the land might have allowed transgression with the formation of tidal flats, and thus decreasing
materials of a clastic origin, have given rise to   condi-
    favorable
tions, for limestone deposition through the instrumentality of
marine organisms, the shells of which make up so large a part
of the Lewistown limestone. But even here, though limestone
preponderates, there are intercalations of limy shale and of
sandstone.

'See Staunton Polio. U. S. Geol. Survey. 1894.

This hypothesis, general as it is, seems to be for the most part, in harmony with the stratigraphy, but it is a matter of doubt whether the ferruginous coating of sand grains could survive the long carriage which it demands. The sharp change from white to r d sandstone seems to indicate that the red color must have been original with the component material for secondary infiliation would, in all likelihood, have caused permeation of the subjacent beds of the older formation. Regarding this matter, then, ther seems to be some conflict of evidence.

MONTEREY PERIOD. During the formation of the great sandstone and the red sands and clays which followed it, the land had been in large part cleared of the residues resulting from earlier removal of soluble constituents: corrasion was in excess of solution; but before the close of the Rockwood period and while the limestone was being laid down, the reverse was true. However in these terranes we have good evidence — in their variformity — of the existence of a coastal-plain. The conditions were essentially those outlined on a pr vious page; land derived material was sorted and separated, the fine particles were carried away and the coarser portions left to ac-

cumulate.

Prepared, then, in a manner similar to the Massanutton materials, the final distribution of the Monterey sediments, as in the previous case, was the result of an upward earth-movement.

The sharp contact of the conglomerate Monterey sandstone with the upper shales of the Lewistown formation is evidence of very energetic currents and therefore, presumably of a very shallow sea. Scouring currents became so efficient that any transition beds, which may have been originally laid down, were entirely removed, and before deposition was again permitted a very large amount of sand and gravel must have been swept across the shallow-water zone and out to the sea.

A period of rest following the initial uplift allowed the accumulation of the sandstone, but how great a thickness it may have reached cannot be determined for the general elevation was resumed and continued until a very large area of the for-

With this erosion the origin of the manganese ores of Little Fort valley is, with little doubt, to be connected, but the data at hand will not warrant the discussion of this problem in the present paper.

ROMLEY-JENNINGS PERIOD. The shales whic rest upon the eroded surface of the Monterey sandstone are extremely fine in grain and in their lower part, highly charged with carbonaceous matter. The conditions under which such sediments could have been the first to cover a recently submerged area are obscure, since the advancing sea must have found the land surface covered with loose material derived from the sandstone. The contact between the two series is, however, imperfectly known and it is possible that the sharp contacts are located on the higher areas of sandstone from which all the debris had been swept into the neighboring hollows. Similarly it is supposable that, with rapid depression, all the coarse material which would ordinarily have been carried out to sea, found lodgment in topographic depressions near shore, and by the time these were filled sufficiently to llow of anything like regular flow of undertow currents the water had become too deep to allow the carriage of any but the finest

sediment. The thickness of the shales (over 1200 feet) shows
that erosion in early Devonian time must have been prolonged
if not active, while their homogeneity is evidence that the
Romney-Jennings period was one of comparative quiet. The depression which resulted in renewed deposition, though of wide
extent, was probably not great in amount, but it must have
been sufficient to restore the coast to the vicinity of its
position before the Monterey uplift.

LATER PALEOZOIC PERIODS. Deformation such as the rocks of
the Massanutten region have suffered is generally conceded
to be possible only under a great load of superincumbent
strata, and this is the sole evidence now remaining that deposition did not cease with the middle Devonian shales. For
the complete folding without fracture which is observed in the
Massanutten sandstone it must have been buried under 10,000
feet of younger formations.[1] The present covering is not in
excess of 3000 feet, and the remaining 7000 feet would carry
the Massanutten section well into the Carboniferous system.

---

[1]B. Willis, Mechanics of Appalachian Structure 13th Am.
Rept. U. S. Geol. Survey 1692 p. 270.

RÉSUMÉ. The foregoing speculations regarding the success
of geologic events during the deposition of the various sediments of the Massanutten Mountain have been mainly confined
to the recognition of changes in the relations of land and
sea, and the deformations by which these changes were brought
about.   In the accompanying diagram an attempt has been made
to represent this interpretation of depositional history in a
concise form.

The diagram consists of three plane curves.   The first
indicates the variable position during geological time, of th
intersection of an assumed radius of the earth with the land-
surface, and, being applicable to the intersection of any
radius with the land-surface may be considered as representa-
tive of changes in the elevation of the land.   Varying depth
of the sea is represented in a similar manner by a curve re-
ferring to the intersection of a chosen radius with the botto
of the ocean.   Since the height of the land above the sea de
termines the activity of erosion, and the depth of the sea
largely controls distribution of sediments, sea-level may be
considered as constant and taken as a plane of reference in

both these curves. The third curve represents the advance and retreat of the shore-line attendant upon elevation and depression. The two elements which enter into both the land and the sea curves are time and vertical displacement, while those determining the shore curve are time and horizontal displacement.

In plotting these curves the columnar section is first constructed and its long direction taken to represent the time element. — It is evident, of course, that no idea of actual or even of relative time-lapse can be conveyed in this manner but this is not requisite; for it is only desired to show synchronism of events, and this could not be better accomplished. — Horizontal distances are represented by the second dimension in the plane of the paper, and vertical distances by the third dimension normal to the paper. Points on the assumed radii at the level of the sea appear in the diagram as lines (ab and áb́) parallel to the time ordinate. The position of the Massanutten area is similarly represented by the line cd. The land curve is placed on the right, the sea curve on the left, with that of the shore in its logical position between them.

75

The diagram presents at once the variations of sedimentation in the different geological periods, and the succession
of earth-movements to which they were related. Lowerin of
the land by denundation, and advance of the sea by marine planation, also appear in the appropriate curves. The land and
sea curves taken together indicate the nature of each movement, whether depression, uplift or tilting and with the aid
of the third it is possible to designate the landward or sea-
ward position of the axis of tilting. Thus shoreward tilting
about a land axis is accompanied by retreat of the sea, but
the same about a marine axis, admits of transgression.

2. DEFORMATION.

POST-CARBONIFEROUS FOLDING. The great earth movements which affected the whole of the Appalachian province after Paleozoic deposition was completed, were far in excess of any which have left records in the Paleozoic sediments. Deforming forces of stupendous power were manifested in a direction perpendicular to the course of the present mountain ranges, and under their lateral thrust, the rocks were first compressed, then crushed or folded, (as they were lightly or deeply buried) until the earth's crust in the zone of deformation was greatly shortened.

The amount of compression in the Appalachians has been variously estimated at from one fourth to one third. In the Massanutten Mountain it has been calculated from the flexures of the Massanutten sandstone and found to be about thirteen per cent.

recently presented by some of our best geologists. Suffice
it to say that when the compressing forces were applied, the
rocks now exposed in the Massanutten region were in the zone
of folding without fracture.

ORIGINAL SYNCLINES. In taking up the investigation of
the geology of Massanutten Mountain it was hoped that it would
be possible to test the applicability of the hypothesis of
original deposition synclines in one of the localities suggeste
as illustrative.' This has not, however, been realized with
any degree of satisfaction.

regions.

Upon the recognition, in the field, of the varying depth of the great fold it became at once apparent, if there were any close relation between the structure of the fold and the thickness of the Martinsburg shales, the latter should become notably thinner toward the ends of the several structural basins.

Accurate measurement of the Martinsburg shale has not been possible but the most reliable estimates show that it has a thickness which cannot be less than 2500 feet and is probably nearer 3000 feet. It thus becomes assured that there is a greater body of the Martinsburg shale in the Massanutten syncline than along the western side of the Great Valley: at Brocks gap[1] Mr. N. H. Darton has found the Martinsburg formation to have a thickness of 1100 feet while on the Staunton sheet they are from 1000 to 1200 feet. Variation in thickness along the axis of the fold has not been proved since all measures of the shale are too indefinite for comparison.

_____

[1] Personal communication.

[1] Staunton Folio U. S. Geol. Survey. 1894.

3. DENUDATION.

Under the heading of denudation it is proposed to follow out, in a very general way, the post Paleozoic history of the region which forms the subject of this paper.

EARLY MESOZOIC PERIOD. In the interval between the close of Paleozoic sedimentation in the Appalachian province and the formation of the early Mesozoic rocks of the Atlantic slope there were great changes in the topography of the Eastern United States. Whereas the drainage had been western into an interior sea the earliest Mesozoic records which we have show that most of the rivers were then flowing to the east. The position of land and sea had been reversed. Any discussion of this interval in this connection would be entirely out of place, but it was a period of great importance in determining the physical features of the eastern part of our continent.

CRETACEOUS PERIOD. The general and long continued erosion which, by the close of Cretaceous time had reduced the continent, raised at the beginning of Mesozoic time, to a base level-surface, has been recognized by remnants of the pene-

plain then produced, which may be traced to the shores of the
Cretaceous sea and there found to merge with the plain formed
by marine deposition.   The topographic development of Norther
Virginia has been parallel with that of the Atlantic slope as
a whole and there is no doubt that a record of Cretaceous base
leveling has been reserved in the summits of Massanutten
Mountain.

It has been suggested on a previous page, that there is
some evidence in the double level of the ridges and peaks,
of two long periods of denudation.   If there are two base
level-surfaces, the lower must be of Cretaceous age,(the Cretaceous peneplain of Prof. Davis), while the date of the upper
cannot be fixed.   If the lower corresponds, however, to the
Severn formation, it is not improbable that the upper is to be
correlated with the Potomac formation.

TERTIARY PERIOD.   The Cretaceous was closed by uplift,
and the surface of low relief subjected again to energetic
erosion, but under conditions where solution was more efficien
than corrasion.   The natural result of this was the speedy
removal of limestones and shales.   Valleys were excavated and
low mountains left in relief.   The reduction of the more

soluble rocks was very complete, and the difference in elevation between the Shenandoah peneplain and that of the Valley ridges shows that the uplift was in the neighborhood of 1400 feet.   Probably it was interrupted and consisted of several minor movements one of which is recorded in the frequent hills and ridges which rise from 400 to 600 feet above the floor of the Great Valley.

PLEISTOCENE PERIOD.   Before the gravels were deposited the rivers had been revived and had lowered their channels in the Shenandoah peneplain, but before erosion had proceeded very far, the competency of the larger streams was in some way diminished to such a degree that they were no longer able to carry their accustomed load.   At this time the gravels were deposited.

Whether or not the reduction of the stream-grade was the result of a general subsidence,[1] can be determined when the presence or absence of the gravels in the northern part of the

---

[1]See Geology of the Coosa River by Hayes and Campbell. (Bul. Geol. Soc. Am. vol. 5, 1894. p.) where similar gravels are thus explained.

Shenandoah valley, is ascertained. It is the opinion of the
writer that fuller investigation will show the phenomenon to
have been connected with the warping which took place after
the later Tertiary base level.

The trend of Massanutten Mountain is at right angles with
the course of the streams which affected the Cretaceous denudation. Hence no great difference of elevation would be expect-
ed in the different mountain-ridges, so that it is evident that
there has been tilting, since the southern end of the Mountain
is 800 feet above the northern end. The Shenandoah base level-
-plain has also an equal slope between Strasburg and Harrisonburg, while the two principal streams have a fall of about 300
feet in the same distance. The existence of such grades in
rivers so voluminous, working in rocks so homogeneous and non-
resistant, is conclusive proof that the deformation has been
very recent.

The diagram of river slopes given on page 47, shows that
both rivers have steep grades between the 500 and 700 foot contour, while between the 700 and 900 marks, the declivity is
only a little more than half as great.

Mr. Keith has shown that the points of equal altitude
upon the two branches of the Shenandoah river being located
nearly upon a north and south line, indicates that the axis of
deformation has the same direction,[1] but the distribution of
river-slopes is sufficient to show that the earth-movement
has been more complete than the simple tilting which he sug-
gests.

---

[1] Geology of Catoctin Belt. p. 373 - 376.

SUMMARY.

The general conclusions which have b
paper are: —

1.   After the deposition of the Cam
stone, a land area was elevated opposite
with its seaward boundary in the vicinity
Ridge.

2.   Subsequent to this early revolut
oscillations of the shore-line, resultant
elevation and subsidence, out the average
coast was not greatly changed from the
ed.   It was now on one side and now on t
tinsburg shore.

3.   The Massanutten syncline marks
-shore zone of maximum deposition, and is,
tive of the hypothesis of original syncli

4.   The general post-Carboniferous
achian province was shared by the Massanu

5.   Since Paleozoic time the region has been several
times elevated, suffering, during the intervals between the
uplifts, more or less complete degradation.   At least three,
and possibly four such uplifts are to be recognized in remnants o baselevel-surfaces.

Arthur Coe Spencer.

Baltimore,Apr 20th,1896.

LIFE.

Arthur Coe Spencer was born at Carmel, N. Y. Sept. 27,
1871. His early education was received from private instruc-
tors and Brook's Military Academy, Cleveland, Ohio. In 1888
he entered the Case School of Applied Science at Cleveland,
from which institution he was graduated in 1892 with the degree of Bachelor of Science. The following Academic year
he was in residence at Johns Hopkins University where he studied Mineralogy and Petrography under the late Professor George
H. Williams, and Paleontology under Professor Wm. Bullock
Clark. During the year 1893-94 he was employed with the Iowa
Geological Survey.

In the autumn of 1894 he returned to Johns Hopkins University and has since that time attended the lectures of Professor Clark upon Paleontology and Historical Geology, those
of Dr. E. B. Mathews upon Petrography, and those of Messrs.
G. K. Gilbert and Bailey Willis upon Physiographic and Stratigraphic Geology. General Biology was also studied under Dr.
E. A. Andrews.

He now holds the University Scholarship in Geology.

www.ingramcontent.com/pod-product-compliance
Lightning Source LLC
Chambersburg PA
CBHW021802190326
41518CB00007B/417